Jan Conrad

Flexible User
Interface - FLUSI

Jan Conrad

Flexible User Interface - FLUSI

An architecture for a context-aware user interface

VDM Verlag Dr. Müller

Bibliographic information by the German National Library: The German National Library lists this publication at the German National Bibliography; detailed bibliographic information is available on the Internet at http://dnb.d-nb.de.

Copyright © 2006 VDM Verlag Dr. Müller e. K. und Lizenzgeber
All rights reserved. Saarbrücken 2006
Contact: info@vdm-buchverlag.de
Cover image: www.photocase.de
Produced by: Lightning Source Inc., La Vergne, Tennessee/USA
            Lightning Source UK Ltd., Milton Keynes, UK

Bibliografische Information der Deutschen Nationalbibliothek: Die Deutsche Nationalbibliothek verzeichnet diese Publikation in der Deutschen Nationalbibliografie; detaillierte bibliografische Daten sind im Internet über http://dnb.d-nb.de abrufbar.

Copyright © 2006 VDM Verlag Dr. Müller e. K. und Lizenzgeber
Alle Rechte vorbehalten. Saarbrücken 2006
Kontakt: info@vdm-buchverlag.de
Coverbild: www.photocase.de
Herstellung: Lightning Source Inc., La Vergne, Tennessee/USA
            Lightning Source UK Ltd., Milton Keynes, UK

ISBN-10: 3-86550-914-2
ISBN-13: 978-3-86550-914-7

Table of Contents

List of figures

List of tables

1 Introduction

1.1 Motivation

For many people mobiles are integrated in the everyday life be-yond telephony. They are regular smart companion. In this co-herence the smart companion is also called *smartphone*.

This special kind of mobile phone is generally considered to be any handheld device that integrates personal information manage-ment and mobile phone capabilities in the same device. It often includes "smart" capabilities, such as PDA functions. The key fea-ture of a smartphone is the possibility to install additional applica-tions to the device. The applications can be developed by the manufacturer of the handheld device, by the operator or by any other 3rd party software developer.

Compared to 2004, smartphones have gained the larger part of the mobile phone market. In a couple of years, it is likely that most of the sold phones will be considered "smart" [1].

By using default applications or installing 3rd party software, smartphones can be utilised for playing movies and sounds, tak-ing, showing and storing photos, organizing schedules, accessing the Internet, route planning etc. The range of possible applications is almost unlimited.

The number of available applications grows persistently. To unburden the user, a preselection of context-relevant application items would be useful. If it's stated that the user is at or near to his office, he needs business-relevant items, for instance the calendar or a ToDo list. If the user is in front of a grocery the purchase list item is relevant. But if this store is closed at the actual time, this item becomes useless.

1.2 Research Questions

The aim is to offer an architecture for location-based user interfaces. The system should work with a minimum of special hardware requirements. Not to overload the user with information, the resulting user interface should be adaptable, context-aware and location-based. The context-data should remain extendible and adaptable. The following research question has been formulated:

How should the architecture be designed to offer a flexible user interface for general smartphones, which is able to adapt to the current context and environment?

1.3 Method

To understand the topic, the background will be analysed and clarified. Relevant issues are introduced and made clear with regard to the research question.

Then, the functionality and features of the prototype and the techniques used are presented.

In an evaluation, the motivation and the research question will be compared with the prototype. Eventually, it will be shown, that the prototype provides an adequate functionality to deal with the research question. At last, extensions and possibilities for further work are presented afterwards.

1.4 Restrictions

The cellular phone networks are moving targets as developments very intense. This means that Symbian OS has to be adapted too. At the origin of this book, Symbian OS 8.0 is the latest available version on cell-phones. It is possible that some used libraries are not available in older or depreciated in future versions.

The user interface in this book is adapted and tested on Series 60 mobile phones. To use the developed prototype on other Symbian OS phones, for instance UIQ, the User Interface needs to be adapted. With regard to this, not so much effort was spent to design a startling user interface design.

2 Location-based services

Three technologies are making progress rapidly in IT: Ubiquitous or Pervasive Computing, mobile Computing and location-based technologies [2].

The software and technologies that will be developed in this project are found in the intersection of these technologies (Figure 1).

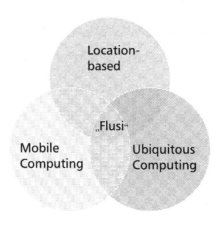

Figure 1: Classification of the topic [2]

Challenging technical issues is part of every design process. A useful and established method in computer science is the principle of "divide and conquer". The key to understand the target is the understanding of the underlying issues. So they are presented first in the following chapter.

2.1 Mobile Computing

2.1.1 An introduction into Symbian O.S.

In 1998, the software developer Psion founded the company Symbian together with the four leading cell phone manufacturers Nokia, Ericsson, Motorola and Panasonic [3]. This company developed – based on Psion's EPOC – the Operating System Symbian OS.

Symbian OS is a robust multi-tasking operating system and is designed for the special constraints of mobile phones such as limited amount of memory and limited speed of CPU.

 It has "…a number of characteristics that make it suitable for mobile computing and better against other operating systems" [4]. Some of the characteristics are:

- **Stability and Security:** A system crash on a mobile phone could be fatal for the user and the acceptance of the phone, so the operating system has to be very stable [3].
- **Electrical Power Consumption:** To save power, the operating system has to save resources [3].
- **Similarity to a Real-Time Operating System:** The telephone features of the device need definable and preferably short response times which are similar to Real-Time Operating Systems [3],[5].
- **Integrated Multi-Mode Mobile Telephony:** Symbian OS integrates the power of computing with mobile telephony,

supporting complex computations, advanced data service and telephony service on a single mobile machine [4].

- **Open Application Environment**: Applications written in different programming languages, like C++ and Java, can be deployed onto Symbian OS [4].

- **Open Standards and Interoperability**: Symbian OS provides a core set of APIs that is shared by all Symbian OS phone. Thus, a program written using the core set of APIs can be used in many Symbian OS phones without the need of modifying the source and recompiling [4].

- **Multi-tasking**: Symbian OS fully implements multi-tasking. Its special feature *active object* allows program requires multi-threading to be written with only one thread with several active objects [4].

- **Fully Object-oriented**: Symbian OS is designed in using object-oriented techniques. It also allows object oriented programs [4].

- **Flexible User Interface**: The graphical user interface in Symbian OS is flexible; developers can design their own user interface easily. Nokia for example developed its own specific GUI called "Avkon" for its mobile phone [4].

The following description of the components of Symbian OS is mainly based on two sources – [6] and [7].

Symbian OS is dividable into four significant component types (Figure 1): Kernel, application, server and engine. As every modern

operating system, Symbian OS has boundaries between the different software components. These boundaries act as protection
against illicit access to memory, other components or hardware.
Thereby the system gains more stability. The components of the
system are shown in the following Figure 1.

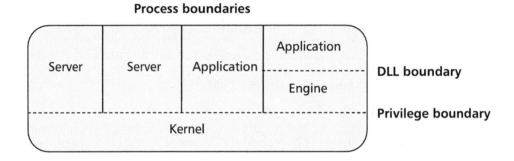

Figure 1: System components

The **kernel** manages the hardware resources of the mobile, such as
the system random access memory (RAM) and hardware devices. It
controls and provides the access to all the other software components of the resources that are managed by it. The kernel has
hardware-supported privilege to gain access to these resources.
Only the kernel has the privilege from the central processing unit
(CPU) to perform certain instructions. The CPU runs other programs – the so-called user-mode programs – without privilege, so
they can only access system resources through the kernel application programming interfaces (API). The boundary between kernel
and user-mode programs is a privilege boundary.

An **application** is a program with a user interface. Each application runs in its own separate process with its own virtual address space, which guarantees that one application cannot accidentally overwrite the data of another application, because their address spaces are entirely separate. The boundary between applications is a process boundary.

A **server** is a program without a user interface. It manages one or more resources providing an API to its clients to gain access to its services. Each server generally runs in its own process that means that the boundary between a server and its clients is also a process boundary. For performance reasons some closely related servers may run in the same process.

An **engine** is the part of the applications that manipulates its data, rather than interacts directly with the user. An application engine may be a separate source module, a separate dynamic link library (DLL), or even a number of separate DLLs. The boundary between engine and application is either a module or a DLL boundary.

DLL and module boundaries that promote system integrity by modularisation and encapsulation are very cheap to cross. The privilege boundary, that promotes system integrity by hiding the kernel and devices from user-mode the code, is more expensive to cross.

In a multi tasking operation system like Symbian OS, each Thread has a stack and a heap memory. The life time of data in the stack is as long as the life time of the allocating function. The de-allocation of data in the heap occurs exclusively by the program-mer.

The most important implementations of Symbian OS are described below.

Descriptors are responsible for string processing in Symbian OS. They offer a safe and consistent work with strings and binary files as well. They are more comfortable than string processing in C and provide full control over used memory.

Descriptors can be distinguished into non-modifiable and modifiable descriptors (Figure 2). Non-modifiable descriptors, marked with a capital letter 'C' at the end of the name of the descriptor class, can be accessed but they are not changeable. They just can be replaced completely. Modifiable descriptors can be accessed and changed. Like non-modifiable descriptors, they contain the current data length information but additionally, they need the maximum length information. This extra information is needed to allow changes within the data. An overview about the descriptor classes is given in Figure 3.

Descriptors can be located in different memory parts. Descriptors in the source code are located in the program memory, i.e. the ROM. Other allocatable memory parts are the stack and the heap (Figure 2). The macro "_LIT" allows the declaration of strings in the source code. So, the command _LIT(KHelloROM, "Hello") generates the descriptor "Hello" in the program memory.

A pointer descriptor is generated by the command TPtrC helloPtr = KHelloROM. "helloPtr" is located in the stack, but the data it points to can be located in the heap or the stack.

A buffer descriptor TBufC lives in the program stack. The command TBufC<5> helloStack (KHelloROM) generates the string "Hello" on the buffer.

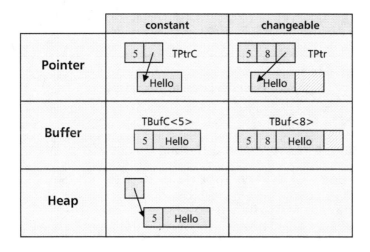

Figure 2: Descriptor overview

In heap descriptors, the data is part of the descriptor object. The descriptor object is located in the heap. To generate a string in the heap, it is necessary to generate a buffer object and copy the data to it: HBufC* helloHeap = KHelloROM().AllocLC. The result is the string "Hello" on the heap. The Pointer helloHeap points at it.

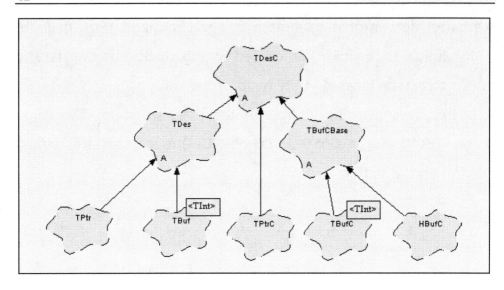

Figure 3: Overview descriptor classes

Memory is a scarce resource on a mobile device. To prevent out-of memory problems, it is very important to free unused memory. Hence, Pointers to allocated memory must be pushed to the **cleanup stack** to handle memory cleanups in leave situations. For instance, it is supposed that CX is a class and x is an object reference. Subsequently, the object is used in a function x->UseL() which causes an error. Now the object reference on the stack disappears and CX on the heap memory is no longer reachable and cannot be destroyed either (Figure 4).

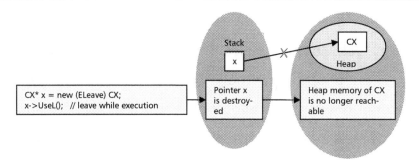

Figure 4: Heap Failure

This problem can be avoided by using the cleanup stack,. After creating the object reference x, a second reference is created on the cleanup stack with the command CleanupStack::PushL(x). Finally, the reference on the cleanup stack is being deleted by CleanupStack::PopAndDestroy(x). So, usually nothing happens except creating and deleting a needless reference. But in case of an error, the function has still a reference on the cleanup stack to delete the object (Figure 5).

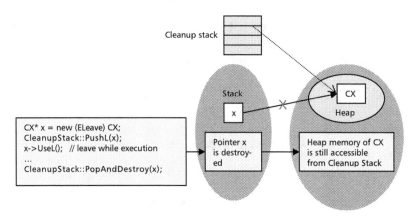

Figure 5: Functionality of the cleanup stack

In spite of the cleanup stack, there are still some situations where memory may leak. To use the cleanup stack, it is necessary to push a pointer to the allocated memory. But this is not possible at object creation in between the allocation performed by *new* and the invocation of an allocation in the C++ constructor. Therefore, the C++ constructor should not fail, which forces a **two-phase construction**. The two-phase construction divides the construction process into two parts or phases. The first phase constructor is the normal construction, denoted by the object's name. This object is pushed onto the cleanup stack. All operations that might leave are done in the second phase construction. So if an operation leaves in the second phase construction, there is still a possibility to deallocate the used memory, because it is already on the cleanup stack.

When a class is inherited, the second phase constructor of the base class has to be called explicitly. It is not called automatically like the first phase C++ constructor. To simplify the use of classes with two-phase construction, it is conventional to provide a static **factory function** NewL(). It performs the two-phase construction and returns a pointer to the created object.

Symbian OS is a multi-tasking operating system as mentioned above. But it does not achieve the multi-tasking capability by multi-threading alone, like many other operating systems. It utilizes an additional mechanism called **active objects**. The main benefit is lower system overhead. With active objects, there exists only one thread per application / server. The requests, sent by one or more active objects, will be scheduled by the **active scheduler**.

So the request can be served in an interleaved manner. Active objects allow non pre-emptive multi-tasking within a single thread. Hence, there is no mutual exclusion code needed to use active objects. In that way, the code complexity is reduced while concurrently achieving multitasking.

2.1.2 Smartphones

The fact that applications can be installed on a cell phone like on PCs give us new opportunities since it is portable and with its means of communication possibilities. Implementing core functionality in software, not in hardware, is also a cost effective approach [8].

Programmable phones were already available in 2001, when Java MIDlets were introduced. But Java always runs on a virtual machine. So machine-oriented properties like access to the Sim-Card, address book, communication interfaces etc. is restricted. Another area which is restricted is the Bluetooth-interface which isn't available at Java MIDP 1.0 and 2.0.

So the presentation of the first Symbian-phone in the end of 2002 was awaited eagerly. The Nokia 7650 and the Sony Ericsson P800 were the first available real smart phones [6].

There is no single feature which is characteristic for a smartphone but usually a phone is considered as "smart" if it offers the following characteristics [6]:

- Colour-display
- Applications like email, calendar, address book etc.
- Extendible through installation of other applications

Symbian O.S is not the only operating system for smartphones but Symbian's share of the market is according to a study of the British market researchers Canalys near 61.4 percent worldwide. The operating system Windows reaches 18.3 percent and PalmSource has 10.5 percent market share (Figure 6). A list of available mobile phones using Symbian O.S. is shown in the appendix.

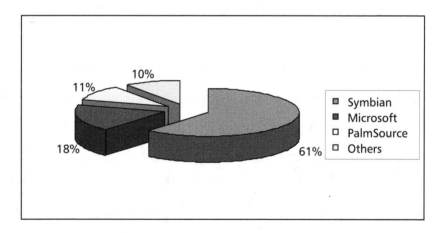

Figure 6: Market Shares of Smartphone's O.S.

2.1.3 Series 60 Platform

Series 60 is a software platform for smartphones with advanced data capabilities. The Series 60 Platform is a source code product that manufacturers can port and integrate into their own hardware designs [9]. Currently there are three editions available (Figure 8).

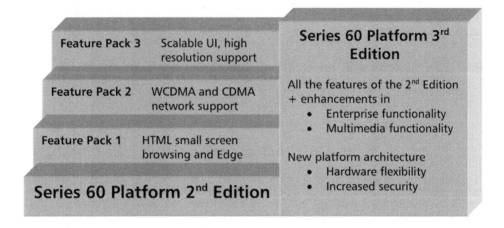

Figure 7: The currently two main versions of Series 60 Platform

The Series 60 Platform 2nd Edition was initially made available in 2003 and was first implemented in the Nokia 6600 device. It has been extended since with three Feature Packs bringing new functionality without compromising the platform compatibility.

The latest edition of Series 60 Platform supports all the features and functions of the 2nd Edition and more - especially in enterprise and multimedia areas. In addition, the Series 60 Platform 3rd Edition introduces a new level of flexibility and security into the platform, enabling manufacturers more easily to create devices targeted to mass markets. By this, Series 60 Platform provides an open, secure and scalable business opportunity to mobile operators and 3rd party developers [9].

2.1.4 Phone-Hardware

Symbian devices consist of a central processing unit (CPU), memory (RAM, ROM), I/O-Devices and a power supply [6]. The hardware components are described in detail below:

- **CPU**: Symbian OS is a 32-bit operating system, so the used CPU are 32-bit-CPUs. Actual smartphones use an ARM respectively StrongARM [10] CPU with between 36 and 156 MHz processing power.
- **Memory**: Smartphones own as memory a ROM and a RAM
  - **ROM**: The ROM contains the operating system and basic applications. The size of the ROM is varying on different Smartphones; most of them have 12 or 16 MB ROM. All files in the ROM are accessible over the drive letter Z:. All files are direct addressable, hence they are executable without loading into the RAM.
  - **RAM**: The RAM acts as working memory for the kernel and applications. Additionally, the RAM provides a persistent storage medium since there's no hard disk drive.

The RAM has a size of, depending on the mobile phone, between 3 - 16 MB.

- **I/O-Devices**: The main I/O-Devices are the display, possibly with touchscreen-functionality and a keyfield. A Card-Slot offer additional memory and is addressed as drive with the capital letter D:. Possible I/O-Devices are also cable, Bluetooth, infrared, WLAN and access over phone-services like GPRS or UMTS.

- **Power supply**: The power supply is eminently important on a battery-powered device. It's normally the embedded rechargeable battery or an external power supply.

2.1.5 User interface design on mobile phones

Are the problems with user interfaces just a fraction of those with desktop-PC UI? Why is it necessary to distinguish between desktop and mobile?

Not only business users use cell pones like in the beginning. Cell phones provide immediately access every time and everywhere to everyone. So it has to be carried everywhere. To reach portability, decreasing size is the easiest way [11]. The battery shall keep the phone alive as long as possible, so the phone-electronic must be optimised for high performance in small size and less energy demand. But the small size limits the number and the dimension of buttons [11]. Also the amount of data, which can be presented on the display, is very limited.

Desktop interfaces are characterized by **direct manipulation**. They provide a permanent representation of all objects of interest, the physical operations that can be performed on them and immediate visual responses to such manipulations [11]. The manufacturer offers a huge set of features, provides them all and the users select his preferences and priorities.

"The main difference in designing UI for desktop environments versus phones is about quantity, desktops can accommodate more." [11]. Small displays with "small" interfaces have to scale downward. They bear on **parallel representation**. Here, the options and items are displayed in a sequential representation. The user has to browse through options, one by one. The hardest point is the user-friendly priorization [11]. Figure 8 shows the path of reasoning used by Nokia, to find the user needs and present them in compact and portable design.

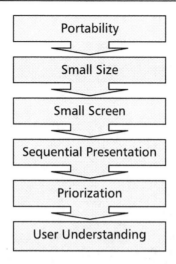

Figure 8: The path of reasoning from portability to user understanding [11]

2.2 Pervasive / Ubiquitous Computing

The history of computing is often divided into three waves [12] (Figure 9). The first wave of computing was dominated by many people serve one computer. The second wave was dominated by personal computers and the most increase wave is the assumption that one person serves many computers. In long term, computing access will be everywhere [13]. But it turned out that not necessarily the number of used computers arises. In fact, by rising portability the availability of processing capability surrounding the user is increasing. But what is ubiquitous and pervasive computing?

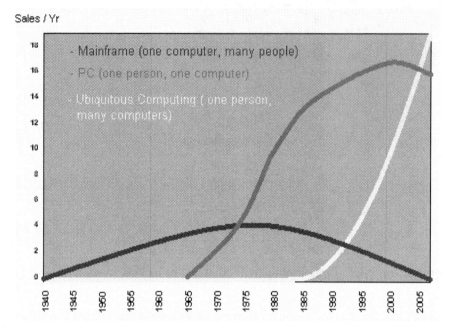

Sales / Yr

- Mainframe (one computer, many people)
- PC (one person, one computer)
- Ubiquitous Computing (one person,
  many computers)

Figure 9: Waves of computing [12]

"A few years ago I found myself on a stage at the MIT Media Lab, arguing with Nicholas Negroponte in front of 700 people. Nick was rhapsodizing about a world in which computerized "intelligent agents" will answer our every need. To illustrate Nick's idea, an actor dressed as a butler introduced speakers and entertained the audience with snide remarks. The butler was fun, up to a point, but also distracting and intrusive. Fortunately, Nick was wrong about what to expect from the third wave in computing. The defining words will not be "intelligent" or "agent", but rather "invisible" and "calm" and "connection"."

by Mark Weiser
Principal Scientist Xerox PARC
March 1996

Pervasive computing is just another term for ubiquitous computing; it is rather used by industry. The term ubiquitous computing was coined by the above cited Mark Weiser. Ubiquitous computing describes the phenomenon of interacting with artefacts and environments with processing and communication capabilities. Computing is integrated in the environment in contradiction to having computers which are distinct objects. Ubiquitous computing is not virtual reality and not a personal intelligent agent [13]. A hard challenge is the integration of ubiquitous computing in the everyday physical world. It has as its goal to augment the real world. It should be calm, almost invisible and in the background [13] so that the user spend his time for the task, not for operating a computer.

The smartphone gives a shape to these abstract ideas [14]. It provides the often postulated cheap wireless networking [13]. It is the personified abstract handheld that are so often mentioned in visions of pervasive computing [14]. "Smart Phone Technology is a result of the convergence of cell phones and PDAs and is steadily becoming ubiquitous..." [14].

In the same way, also software is moving from desktop computer into new and unknown territories [8]. This entails many new challenges. Deployment and maintenance on ubiquitous devices is different to deploying to and maintaining desktop PCs [8]. Also a tight integration of services is needed, as demanded by the small form factor of the devices typically found in ubiquitous computing environments [15].

But ubiquitous computing is not without issues. In connection with pervasive / ubiquitous computing, five critical topics are currently discussed [16]:

- Privacy
- Security
- Unmastered Complexity
- Free Speech
- Intellectual Property

2.3 Location-based services

The development of technologies to communicate with mobile devices and sensors forges ahead. By the advances in wireless communication, and positioning technologies, location-based services become possible. Many new communication innovations are on the brink of breakthrough. But what are location-based services? This chapter will set the main focus on location dependent services, the term context awareness will although be reflected.

Definition:

A Location-based service is a *"...service, query or process whose return or other property is dependent on the location of the client requesting the service and / or of some other thing, object or person."* (ISO TC 211)

The term location-based service denotes applications that connect geographic locations with the general notion of a service [17] to add value to a user. According to Vodafone, location-based services are distinguishable into two types:

- **Active services:** The User initiates the location request (e.g. find-nearest... applications)
- **Passive services:** A third party locates the position of a location-based service participant (e.g. find-a-friend functionality)

2.3.1 Entities

The world consists of infinite many things. In the context of this book, we have to care about the resources and items of a mobile phone. To handle them, it's necessary to map the features, applications and possibilities of the target devices to a finite set. This set is called *universe* and consists of representations of "real-world things". The elements of the universe are called *entities*. To iden-

tify them, an entity needs a set of properties that makes it identifi-
able [18]. They also have to be distinct [18].

2.3.2 Locations

We consider that a location is a place or range, where an entity is
positioned in the real world. Locations can be expressed in many
ways using different frames, for instance as an absolute spatial location, a descriptive location or a relative location [19]. They can
be distinguished in physical positions like 49° 39′ 31″ or in symbolic locations like home or workplace [20]. Physical and symbolic
locations are identified by coordinates. A *coordinate* is an identifier, which specifies the position of an object with respect to a
given coordinate system [21]. A *coordinate system* adds a special
context to a set of coordinates. It's possible to classify coordinates
of locations in two different formats:

- **Geometric coordinates**: Positions are defined in form of tu-
  ples relative to a reference coordinate system [21]. Geometric
  coordinates can be distinguished further in:
    - Absolute coordinates: These coordinate systems are
      valid worldwide like for example the WGS84 system or
      the Transversal Mercator System. It adds to every point
      of this planet a triple of coordinates.
    - Relative coordinates: Geometric coordinates can not
      only refer to global areas, they can refer to local areas

too e.g. Cartesian coordinates in a room with indoor navigation. This coordinates can also describe a self-centred coordinate system, all other locations are described with respect to the own position.

- **Symbolic coordinates**: Positions are defined by abstract symbols, for instance by Cell Ids or Ids of access points the device is connected to. The Symbolic coordinates describe only the affiliation to a sector.

2.3.3 Location sensing techniques

Many systems have treated over the years the problem of location sensing [20]. There exists many approaches but each of them solves slightly different problems. The location is a very personal and unique information. So it is a superb attribute to use in personalization [5] and a significant context parameter [22]. Almost all information is relatable to a location. To obtain the position, a positioning system is needed. According to Hightower and Borriello [20] we can choose from three major techniques:

- **Triangulation:** Triangulation uses the geometric properties of triangles to compute the current location. It is divisible in Lateration and Angulation. Lateration calculates the location by measuring the distance from multiple reference positions.

Angulation is similar to Lateration but instead of distances, the angles of the reference positions are used [23].

- **Proximity:** A proximity location sensing technique entails determining when an object is "near" a known location [23].

- **Scene analysis:** The scene analysis location sensing technique determines the location by observing a scene from a particular vantage point [23]. To obtain characteristic features of the scene, it will usually be simplified.

No location sensing technique takes perfect care or works well in all situations [24]. More accurate positioning requires more resources and more technical effort. So it's important to choose the positioning technique as accurate as necessary and as easy as possible. Ericsson researched into this issue and developed at high level a scheme for the level of accuracy required (Table 1). An overview of state-of-the-art positioning techniques is in the appendix.

Application	Accuracy	Application	Accuracy
News	Low	Gaming	Medium
Directions	High	M-Commerce	Medium to High
Traffic Informa-tion	Low	Emergency	High
Point of Interest	Medium to High	Sensitive Goods Transportation	High
Yellow Pages	Medium to Low	Child Tracking	Medium to High
Car Navigation	Medium to High	Pet Tracking	Medium to High
Personal Naviga-tion	High	Electronic Toll Collection	Medium to High
Directory Assis-tance	Medium to High	Public Manage-ment System	Medium to High
Fleet Manage-ment	Low	Remote Work-force Manage-ment	Low
Car Tracking	Medium to High	Local Advertise-ment	Medium to High
Asset Tracking	High	Location-Sensitive Billing	Medium to Low

Table 1: Required level of accuracy [17]

2.3.4 Context awareness

The term context embodies about everything that constitutes the user's environment and the users themselves.

An existing example for context awareness on desktop-PCs is ICQ. The ICQ-Messenger assumes after a period of inactivity, that the user is away and switches automatically its user mode to "away". On mobile devices there is much more context information available because the device is always on and always with the user [22].

2.3.5 Filters

The intention is the customisation of services to the context and the user profile. Location based services process the movement and queries of their users in multidimensional databases, i.e. data warehouses [26]. The dimensions usual are:

- Location of the user
- User profile
- Time

This context information also known as presence information refers to the physical and social situation, the computational devices are embedded [22].

2.3.6 Privacy and Security

Though this is not part of this paper, it should be mentioned that location-based services also evoke some perils. Location based services and the connected personal information, for instance positions, preferences and user profiles are a high privacy risk. Also the same spamming problem as in email traffic is possible. As result of these potential threats, European Parliament has incorporated these issues in its *Directive on Privacy and Electronic Communications* [5].

3 Flexible user interface prototype "Flusi"

3.1 Scenario

The scenarios described below are based on every day situations. The Flexible user interface prototype Flusi is used to make these situations easier. Services that do not benefit from positioning and personalisation are not described in detail. These uses-cases should help illustrating the benefits of using the mobile service.

Peter is in his office. With regard to the current cell id, the mobile recognize that it is in or near Peter's office. Here, he needs primarily his phonebook, the calendar item, a Bluetooth connection for synchronization with his desktop PC, the voice recorder and his ToDo list. So these items are displayed. Other items like the notepad, browser or the email-client are now useless because his PC offers a more comfortable way to do these tasks. Therefore these items are hidden. Also leisure time features like camera, camcorder, SMS and MMS are not displayed. By this preselection, Peter can find relevant features easily and comfortably without tedious clicking and scrolling (Figure 10).

Figure 10: Office relevant items, filtered by cell id

In the afternoon, Peter leaves his office for an external business appointment. It's an unusual place for business, so Peter has to select the context manually (Figure 11). He browses through the points of interest and select the item "Business Trip". On a business trip, other items are relevant than in office. Peter doesn't need the Bluetooth connectivity for synchronization anymore. He does not have a desktop PC, so he needs his mobile to write emails and notes. Hence accordant items appear.

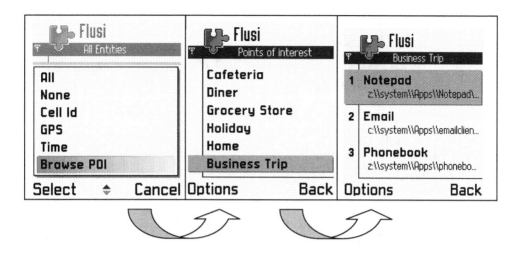

Figure 11: Selecting POI manually

Homebound, Peter switches on his GPS receiver to enjoy more exact functions of his location based service. Peter continues to a grocery. At home, he completed an electronic shopping list. Now, the mobile phone recognizes the proximity to the grocery and the grocery list item appears with a "beep" on the display (Figure 12). Peter is reminded and finds the shopping list application very comfortable in the list of the relevant items on the display of his mobile. Afterwards, Peter continues his walk and goes along a drugstore. He has some sanitary products on his shopping list too, but this drugstore is only open till 5 p.m. and it's already 6 p.m. So there's no need to inform the user about this store and the mobile doesn't display the shopping list item near this location.

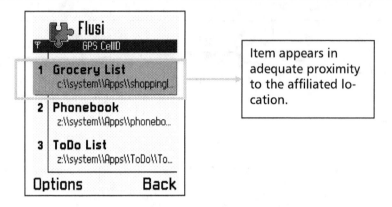

Figure 12: Selection of Locations via GPS

One week later, Peter is on holiday in London. Of course, he brings his mobile phone. By the cell-id, the mobile phone recognizes the location. The schedule and all work relevant stuff are automatically not displayed. In this way, Peter can really relax on his holidays. Instead of postcards, he sends MMS and SMS messages to his friends. These holiday-relevant items are of course displayed. The most important things on his trip are the built-in mega pixel digital camera and camcorder functionality of his mobile. To get more exact locations Peter switches on his GPS-Receiver. When the mobile detects a dignified place for a picture, it switches the camera icon on and notices this with a "beep". So, Peter can enjoy his trip and doesn't miss any sight. Another important tool is the currency converter which is displayed in the UI. The clock-application appears automatically from 10 to 12 pm. This is Peter's usual time period to go to bed and he didn't take any additional clock with him.

3.2 Architecture

The system consists of three main parts (see Figure 13):

- - The User Equipment
- - The Provisioning Server
- - The Positioning part

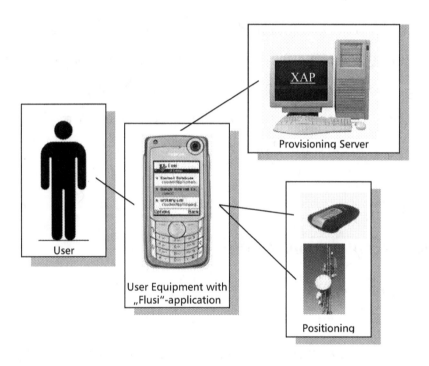

Figure 13: Architecture of Flusi

The User Equipment is the device, on which the Flusi-client is running. It acts as the interface between user and system and overlaps the built-in user interface. By selecting an item in Flusi, the underlying application is executed. After closing the application, Flusi becomes active again.

The Provisioning Server offers the database and points of interest. It is divided as a separate part to save phone's resources and offer accessing possibilities for 3rd party. The same points of interest are accessible for several users.

The Positioning part includes all positioning devices usable for location-based services with a mobile phone. The GPS-Receiver is the most descriptive positioning device because it's a separate tangible thing. Positioning over cell id doesn't result from a single external device. Due to the fact that data from external devices are necessary, it is also counted among a self-part.

3.3 Used Techniques and Functionality

3.3.1 The universe

The universe of the mobile phone is stored in a XML-file (Figure 14). The describing DTD-File can be found in appendix. This file contains the entity name, the entity unified resource identifier <URI> and a label for the display. This file is device-specific and adapted to the current system. The distinction in entity name, URI and entity label is meaningful for various reasons. The URI is device-specific, for example *"z:\\system\\Apps\\browser\\browser.app"*

is the URI for the Browser on the used Nokia 6680. On the phone, the URI is unique. The entity name is a device independent short-cut for that URI to ease the handling. It has to be unique on the device, but also at the entire Flusi system. The entity label <ELabel> is responsible for a user-friendly display text. Because of the constraint of uniqueness of entity names, the <ELabel> allows a more flexible nomenclature. They don't have to be distinct and it can be useful to use the same display text for some different entities. The XML-architecture allows increments like adding icons to the entity easily by adding additional tags to the XML-file.

```
<Universe>
- <Entity Name="Grocery_List">
       <URI>c:\\system\\Apps\\shoppinglist\\shoppinglist.app</URI>
       <ELabel>Grocery List</ELabel>
       </Entity>
- <Entity Name="Contacts">
       <URI>c:\\system\\Apps\\contacts\\contacts.app</URI>
       <ELabel>Contact Database</ELabel>
       </Entity>
- <Entity Name="Browser">
       <URI>z:\\system\\Apps\\browser\\browser.app</URI>
       <ELabel>Internet Browser</ELabel>
       </Entity>
</Universe>
```

Figure 14: Universe file

3.3.2 Locations and filters

The location data is saved in an XML-File (Figure 15). The describing DTD-File can be found in appendix. Several filters are applicable on locations. So, a location's attribute "determination" destines the convenient filter. The element <locname> adds the location name for the user-interface. <locdata> consists of the filter specific data. An element "entities" lists all relevant items for the location.

```
<locations>
-  <location determination="None">
        <locname>Living Room</locname>
        <locdata>Home</locdata>
        <loc entities="Browser Grocery_List Contacts Sms Mms" />
        </location>
-  <location determination="GPS">
        <locname>Grocery</locname>
        <locdata>478046 432432</locdata>
        <loc entities="Grocery_List" />
        </location>
-  <location determination="Time">
          <locname>Grocery_MainStreet</locname>
          <locdata>10 18</locdata>
          <locconnection>GPS</locconnection>
          <loc entities="Grocery_List" />
    </location>
```

Figure 15: Location file

On the phone, one or more filters can be switched on/off over the menu. Five filters are implemented in the prototype. They are chosen from the menu, enabled filters are displayed in the top of the display (Figure 16). The implemented filters are:

- **All**: This filter deactivates all other filters and all on the phone currently available entities are shown. The term "All" doesn't appear in the XML-File, it's a reset-function for the following determination methods.

- **Cell ID**: The entities are selected convenient to the current cell id. The cell id of the location is saved in the element <locdata>.

- **GPS**: This filter needs external hardware i.e. a GPS-Receiver. By activating that filter, the phone displays the items. The GPS-coordinates are saved in the WGS 84 format in the element <locdata>.

- **Time**: The elements are selected with respect to the current time. In the <locdata> element two numbers are available, a starting hour and an end hour in 24h system.

- **None**: Not all entities fit to one of the filters. To get a possibility to switch these runaways on/off they are catched in this filter.

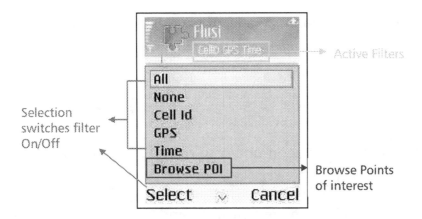

Figure 16: Filter Selection

Is a filter switched to "on", all locations with fitting "determina-tion" and <locdata> are displayed. Exceptions are the filters "All" and "None", they don't need any location data. If more than one filter is activated, the different filters are connected by an *or*. That means if a filter criteria is fulfilled for one or more locations, the entity will be displayed. Thereby, every entity is only displayed once, regardless of how many locations force it. But it can be use-ful to connect locations by *and* and not by *or*. The possibility to do this is given by the element <locconnection> in the XML-file. This element can link one location to another. This connection deals with the intersection of the location entities of both locations. These entities will only be displayed if the <locdata> of the two connected entities are fulfilled.

By selecting the menu item "Browse POI", it's possible to browse all accessible locations in the local location-file. On a location, one or more entities can be eligible, depending on the attributes enti-ties in the loc-tag. This feature enables the selection of locations and circumstances manually and offers a filter independent use of Flusi.

3.3.3 Distinction between universe file and location file

Using different XML-files for universe and locations is reasonable. The universe is device-specific. By offering the possibility to install applications, the directory structure of every Smartphone is different. Applications can be located in different paths, so the URI of the universe file is not general and needs to be adapted.

Locations are not device-specific. They need only the correct name of the entity but not their positions in the file system. So this disjunction allows a device independent location database.

3.3.4 XML-Parser

The eXtensible Markup Language is a standard for man- and machine readable documents. The data is stored in a tree structure. XML provides an easy and flexible way for data exchange between applications. But the used version of Symbian OS doesn't offer a build-in XML-Parser. Flusi features its own XML-Parser. It is based on Java's SAX parser, the Simple API for XML parser. It obtains a XML-file and a fitting processing instruction called handler. While parsing the file, the parser calls, for every tag, different event-handling methods of the associated handler. These methods build the data of the XML-file data.

By using different handlers, the parser is universal and applicable for various XML-file types. In the Flusi-prototype, the entity-file and the location-file are parsed with the same parser but with different handlers.

Another widely-used XML-Parser in Java is the Document Object Model-parser or DOM-parser. This parser creates a tree-structure of the XML-file in the memory. Because of the lack of resources on a mobile phone, this approach is not favourable with the current tasks.

3.3.5 GPS

To obtain a better positioning, Flusi uses a GPS-Receiver with NMEA output. Navstar GPS or simply GPS is an outdoor location sensing technique. It gives "every centimetre of the surface of the planet its own unique address that can be understood by anybody through the use of a universal geocoordinate system." [27].

GPS uses as universal geocoordinate system the WGS 84 System. To describe the surface of the earth, this system uses an equipotential or Gravimetric surface model, a so called Geoid. The WGS 84 Geoid is one of the latest and best-fit Geoids available.

GPS consists of three main segments:

- Space Segment
- Control Segment
- User Segment

The Space Segment and the Control Segment are operated by the U.S. Military and administrated by the U.S. Space Command and the U.S. Air Force [27].

The Space Segment implies the constellation of satellites that are currently in orbit, including operational, inoperable and backup units. Figure 17 shows an actual GPS Satellite and table 2 gives some technical data.

Physical Characteristics:

Name:	Navstar
Manufacturer:	Rockwell International
Weight:	~900 kg
Size:	~5 meters wide
Lifespan	7.5 years

Orbital Parameters:

Altitude:	~20200 km
Orbital Period	12 hr
Orbital Plane	55° to equatorial plane
Number of pla-nes:	6
Orbital Spacing	60° apart

Figure 17: GPS-Satellite   **Table 2: Technical Data**

The Control Segment maintains the integrity of the satellites and the data they transmit. It consists of a Master Control Station at Falcon Air Force Base in Colorado, five unmanned monitor stations, located strategically around the world and three primary ground antennas, located approximately equidistant around the equator [27] (Figure 18). The Master Control Station sends periodically corrected position and clock-time data to the three ground antennas which upload them to the satellites.

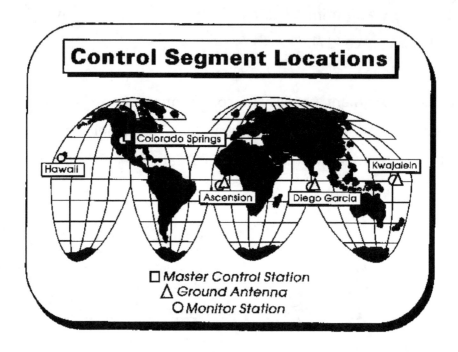

Figure 18: Control Segment [17]

The user segment covers all military and non-military users of GPS. With respect to Flusi, these remarks concentrate on civilian use only. The satellites send at a time mark t_0 on its internal clock a signal to the earth. The receiver generates that signal at t_0 on its internal clock too. When the receiver picks up the signal from the satellite, the time interval Δt of the signal-runtime is determinable by comparing the internal signal and the received signal with respect to the time t (Figure 19).

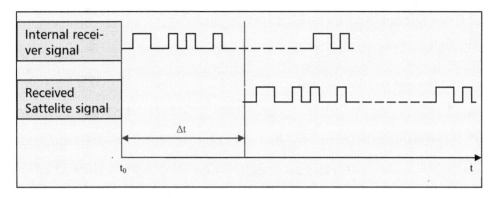

Figure 19: Measurement of signal runtime

The available and necessary data, broadcasted by the satellite are:

- System time and time correction
- Ephemeris data of the broadcasting satellite
- Almanac data (ephemerid data of all satellites)
- System state

If the receiver receives signals from at least three satellites, it could theoretical determine its position with those Δt. But by developing the GPS System, one goal was to make cheap receivers possible. So the receivers are not equipped with very expensive but very exact atomic clocks like the satellites. An atomic clock is accurate to within billionths of a second per month and cost something like $200.000 [27]. But the exact time can be determined too by evaluating the data from a fourth satellite, these are at least four equations with four unknowns. So receivers can be equipped with cheap and very common quartz-clocks. The resulting equation system is not linear, so it's complex to solve the equations and requires some iteration.

3.3.6 GPS-Interface over Bluetooth

In our experimental setup, Flusi uses a Bluetooth GPS-Receiver from SysOn (Figure 20) to receive the GPS data. It sends position data via Bluetooth to the mobile phone. Bluetooth is an industrial specification for wireless personal area networks. It is named after the Danish king Harald Blåtand who is known for his unification of previously warring tribes from Denmark, Norway and Sweden [1]. Bluetooth is designed for low power consumption, with a comparatively short range (in mobile phones about 10 m).

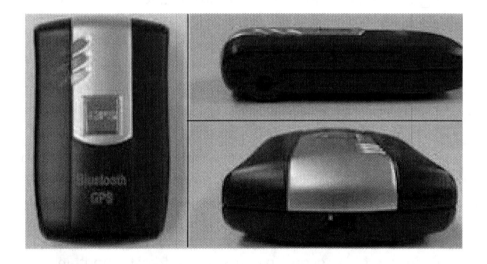

Figure 20 SysOn Bluetooth GPS-Receiver

To connect Flusi with the GPS-Receiver, the user has to select the menu point "Connect to GPS" (Figure 21).

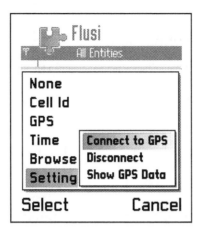

Figure 21: Connect to GPS-Receiver

If the Bluetooth device in the mobile is switched off, a dialog box opens to switch it on. Then a list with all available Bluetooth Devices appears. After selecting the GPS-Receiver, entering the receiver specific Pin-Code, the Bluetooth-Connection is established.

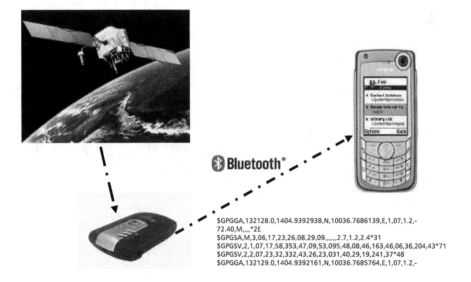

Figure 22: GPS Positioning over Bluetooth

After establishing a connection, the receiver streams the position information (Figure 22). The received data is in NMEA format, the most common format for GPS data exchange. The following seven datasets are very rife:

- **GGA** (GPS Fix Data)
- **GLL** (Geographic Position – Longitude/Latitude)
- **GSA** (GNSS DOP and Active Satellites)
- **GSV** (GNSS Satellites in View)
- **RMC** (Recommended Minimum Specific GNSS Data)
- **VTG** (Course over Ground Speed)
- **ZDA** (Time and Date)

In Flusi, the RMC dataset is used. A typical RMC dataset looks like this:

*$GPRMC,161229.487,A,3723.2475,N,12158.3416,W,0.13,309.62, 120598,,,*10<CR><LF>*

The following table contains the values and description of the dataset above:

Date	Description
$	Start of a dataset
GP	Information from a GPS-Receiver
RMC	Identification of RMC dataset
161229.487	UTC-Time
A	A=data valid or V=data invalid
3723.2475	Latitude
N	N= north or S=south
12158.3416	Longitude
W	E= east or W=west
0.13	Speed Over Ground
309.62	Course Over Ground
120598	Date
*10	Checksum
<CR><LF>	End of message

Table 3: NMEA-Data

If the GPS-Filter is active, Flusi parses the position every second and tests if the actual position is inside a fixed radius around the location position. If the actual position is inside, the entity will be selected.

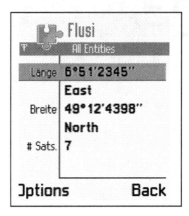

Figure 23: Parsed GPS-Data

The accuracy of the used GPS-Receiver is, according to manufacturer information, about 10m. But in experiments, the reached accuracy was much better. The menu-option "Show GPS Data"
shows the parsed NMEA-message (Figure 23). This feature is for
demonstration purposes only and serves no intention in the Flusi-
architecture.

3.3.7 Provisioning

The location information should be updatable. The mobile phone
needs a device with a large storage space, which is always reachable from everywhere. So Flusi is integrated in the eXtended Application Provisioning (XAP) system from the MSI, Växjö University
and an Apache Tomcat Web server. As database a MySQL database is used.

The data-transfer processes over the HTTP protocol. It allows the required speed and offers the lightness for distributed collaborative hypermedia information systems. It is a generic object-oriented stateless protocol.

In a HTTP protocol transaction, the following lines are set by the client in the RFC822 format headers:

- From
- Accept
- Accept-Encoding
- Accept-Language
- User-Agent
- Referer
- Authorization
- Charge-To
- If-Modified-Since
- Pragma

Important for the XAP system is the line User-Agent. It is set to "Flusi 1.0" to inform the XAP system about the client type.

HTTP supports different types of request methods for different purposes, some of them are: GET, HEAD, POST, PUT, DELETE, LINK, UNLINKE, TRACE...

Flusi uses the GET-method. It's the oldest and easiest method in HTTP. This is useful to save online-traffic and costs.

A possible request could look like this:

GET / HTTP/1.1[CRLF]

Host: http://flusi.homeip.net/xap/gps [CRLF]

User-Agent: Flusi 1.0[CRLF]

Connection: close[CRLF]

Accept: */*[CRLF]

The header from the answer could look like this:

HTTP/1.1 200 OK[CRLF]

Date: Thu, 15 Jul 2005 15:22:22 GMT[CRLF]

Server: Apache/1.3.5[CRLF]

Accept-Ranges: bytes[CRLF]

Content-length: 46[CRLF]

Connection: close[CRLF]

The body from the answer is the locations XML-file (Figure 15). This file is composed in real time by the XAP-System. The GET request calls a Java Servlet. This Servlet creates a SQL query and applies it on the connected MySQL-Database. In the query result, it composes the fitting location data. This data is the body of the HTTP answer. Flusi receives the data and saves it as a XML-file on the phone. A complete sequence is shown in Figure 24.

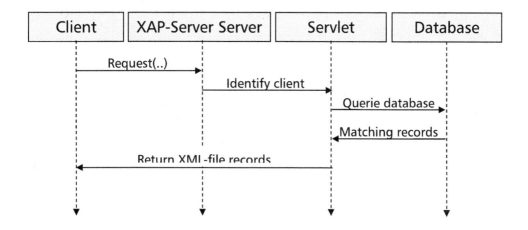

Figure 24: Sequence diagram of a server request

3.3.8 Cell id

Each mobile phone tries to connect to exact one base station, when it is switched on. The region a base station covers is called cell. To avoid uncovered regions, cells are overlapping. At these margins, the phone connects always to the station with the strongest signal. The base station transmits a location area id and a cell identifier id. It is assumable that the nearest base station is the one with the strongest signal. This fact allows for an approximation about the position of the phone (Figure 25).

In a GSM-Network the cell-size is constant. There are maps with cell ids and their geographical regions. Each base station has a fixed capacity, so the size of the cells depends on the population density: In urban regions, the cells are smaller and in countrified regions cells are larger. In 3G networks, the cell-size depends on the number of connected phones, the cell is "breathing". So it's more difficult to link from the id to geographical data. The broad-cast-frequency of the 3G network is with about 2000 MHz higher than those from the GSM-Network (900 MHz / 1800 MHz). This implies a higher damping, hence the 3G-cells have to be smaller than GSM-cells.

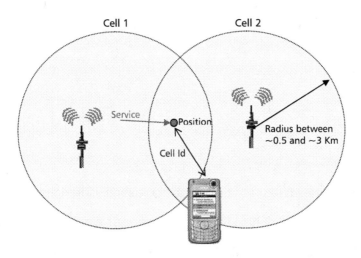

Figure 25: Positioning over Cell-Id

4 Evaluation

The aim of this book is to design an architecture to offer a flexible user interface for general smartphones, which is able to adapt to the current context and environment. The following section measures the presented work against this statement.

4.1 Generality

The target device should be as general and common as possible. Symbian OS and Symbian OS driven Smartphones provide the necessary diffusion rate (like presented in Chapter 2.1.2) together with sufficient processing power, to establish a common standard. Furthermore, these smartphones are open for third party software developers.

A tight integration is needed. Due to the previously described facts a mobile phone has, compared to a Desktop PC, scarce resources. But the larger part of the software should run on the phone [29]. Nevertheless a server should store and provide the relevant data [29]. To solve this, Flusi uses a server with a database, which disencumbers the memory of the phone. This architecture allows additionally an easy data exchange between more than one user. Updates and renewals on the dataset are done once and changed for every device using Flusi. These advantages are not possible without a central server.

There are still some problems with the state-of-the-art devices but most of them could be solved in the near future. In the practical work it turned out that the access to application development for Symbian OS is harder than for example for Java. At the moment, Symbian OS is poorly documented and it's hard to get answer to special questions. Fundamental functionalities are announced but not available yet, e.g. the required XML parser.

The development device is usually not the target device. So, a cross-compiler and deployment facilities are needed and the development takes place on an emulator. This entails behavioural differences between the emulator and target devices such as performance and availability of libraries. Not all phone features are testable on the emulator; for instance Bluetooth functionality is only available at the target device. So, a lot of time is required for testing and debugging these features.

4.2 Flexible user-interface

Product segmentation is an extremely important point in mobile phone development [11]. For instance Nokia has a strict

	Balancer	*Controllers*
CUSTOMER SEGMENT	*Balancer*	*Controllers*
PRODUCT CATEGORY	*Expression*	*Classic*
SOME CURRENT PHONES	Nokia 3330	Nokia 6610
USER INTERFACE ATTRIBUTES	*Intuitiveness Personalization Simplicity*	*Efficiency Productivity Backward compatibility*

Table 4: Nokia's Customer Product and User Interface Segmentation Part I [11]

CUSTOMER SEGMENT	*Experiencers Impressers,*	*Impressers*
PRODUCT CATEGORY	*Fashion*	*Premium*
SOME CURRENT PHONES	Nokia 7210	Nokia 8910
USER INTERFACE ATTRIBUTES	*Individuality, Appearance, Fashionability*	*Exclusivity, Individuality, Attention to details*

Table 5: Nokia's Customer Product and User Interface Segmentation Part II [11]

segmentation to customer segments and user attributes (Table 4: Nokia's Customer Product and User Interface Segmentation Part I [11]Table 4 and Table 5). Thus the user interface is fixed to the target device.

Many manufacturers argue that mobile phones are consumer goods, so they can't expect from users to configure it themselves [11]. But examples like the mobile phone market in Japan show that personalization is a demand. The UI is partially embedded into hardware and adapted to the target segment. But this background restricts the personalization of the device. An approach like Flusi allows a phone – and phone segment – independent personalization and offers a perfectly user-adapted Look-and-Feel or *flexible user-interface*.

The user interface provided by Flusi isn't fixed. The use of the XML to describe the elements allows a flexible framework, where the user interfaced is attached. The list design is an appropriate manner to present the actual items visible. Needed applications are easily and directly selectable. The user defines the circumstances and time when an item is visible.

Because of the habitualness of mobile phones, most mobile phones are bought by replacement users. Through the client server principle in Flusi, older user preferences and settings are directly portable. They are available on the server and directly usable on other devices with installed Flusi application. This adds to the flexibility of the user-interface a flexibility of the target device.

4.3 Adaptability

Location-based services and context awareness include a broad area of topics. This book has its main focus on the relation of user interface and location. Flusi is a so called *active* location-based service. The application considers the context and adapts the set of available features according to the selected filters. The dimensions of data are

- **location**
- **time**
- **user profile**

These are considered by Flusi to identify the context of the actual situation. We believe that these dimensions are significant for the customisation of services to the context. Through them, it is possible to adapt the flexible user-interface to the current context and environment. All filters are selectable by the user. In this way, the user retains control over the considered context.

Positioning is required to adapt the presentation. Navstar GPS is at the moment the best available global positioning system and provides an excellent Lateration framework [20]. The worldwide satellite constellation has ubiquitous coverage and allows a location determination within 5 meters [20]. Perhaps, Navstar GPS will be detached in some years by the more accurate European GALILEO System.

Moreover mobile phones with build-in GPS will be more common. The USA as well as Japan is establishing location sensing as mandatory function for emergency calls. GPS is only available outdoors. Also street canyons or shadowing effects in cities entail impreciseness in positioning. So GPS is not the all-in-one device suitable for every positioning purpose.

The cell-id allows approximately localisation without any special hardware. No additional hardware entails any additional energy consumption and longer durability. The cell-id positioning technique is available everywhere, where service is available. Unfortunately, this technique is quite imprecise. Only an estimation of the current position is possible. The differences in cell-size are an additional unknown.

"Hand operated positioning" through selecting the context manually is not an option of the user interface, but this is not a restriction. Lest we forget that this is the "normal" and traditional way of running a mobile phone interface.

Recapitulating, it's safe to say that there is no available positioning technique fitting for every situation today. The correct choice of the positioning technique is dependent on many conditions. Flusi provides a set of methods that covers a wide range of circumstances. The user is responsible to select the positioning technique that works best and fits to the user's intention.

4.4 Flexibel user interface prototype „Flusi"

The context aware approach offers a new and powerful alternative to the less comfortable interface provided off-the-shelf. The, by manufacturer forced, priorization of items becomes context- and user-dependent and dynamically adaptable.

Also the possibility to prepare the data by third party, which maybe have commercial interests, is a powerful engine to press location-based services ahead. 3G and UMTS are expected to offer a powerful environment for location-based services. The capability of UMTS goes far beyond those from 2G [28].

The use of mobile phones allows a calm integration in the physical world, like demanded in ubiquitous computing. The mobile phone is small and almost independent from wires. It augments the real world by many useful workday features. A context aware application like Flusi yields smartphones a further step closer to the utopistic ideal of ubiquitous computers.

So, the prototype Flusi offers an architecture for a flexible user interface. The used devices are very general and common. Its context awareness offers an adaptation to the current environment.

5   Conclusions and Future Work

5.1 Enhanced Provisioning

The XAP system is more powerful than what is utilized in Flusi. It offers lots of possibilities to customize location data. The XAP-System can be expanded to offer the Flusi installation file for many devices. The device connects to the server and XAP returns the fitting application. This affords an easy and wide automated distribution of the application. Features like Bluetooth interface, infrared interface or WLAN support can be added or removed, fitting the current device. Also points of interest can be selected compatible to the connected devices. XAP could identify automatically the client's position and the feature set. So, it is possible to save data traffic and device memory.

5.2 Data integration

A special issue is the administration of locations. The location database should be extensible by:

- Each user
- Commercial third party
- Non-Commercial third party

Each user should be able to enter new locations and new relations; in ideal case even everywhere and every time (Figure 26). Therefore different gateways for database-access are supposable. The most comfortable is definitely a Web interface. It's reachable over desktop PCs as well as via mobile phones and all other HTML enabled devices. But also special applications to maintain the databases are imaginable. The MySQL database offers the flexibility to use different techniques parallel. Commercial third parties, like supermarkets and stores, can integrate their shops to acquire customers. Also customers gain more comfort.

Moreover, non commercial third party benefit from adding points of interest. A city which bid for tourists can integrate important sightseeing points and enhance visits. The mobile phone becomes a personal tourist guide.

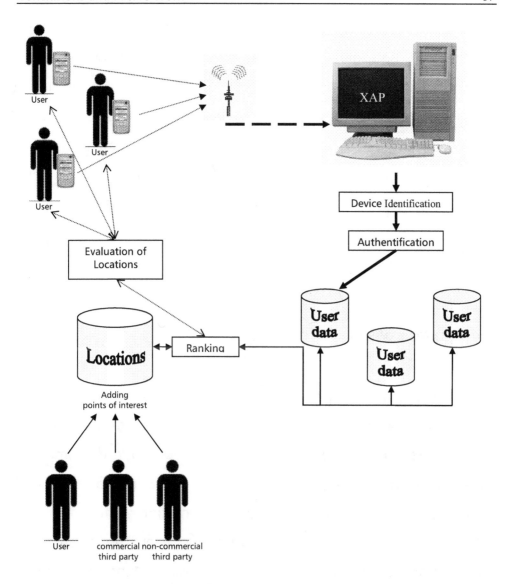

Figure 26: Expansion of Flusi prototype

5.3 Variable radius of relevance

Another aspect is the radius where a location is relevant for the user. In Flusi, this radius is hard-coded. But to enhance the features the radius could be made flexible and dependent on different factors. The radius can be added in the location XML-file. The radius can also be conditioned by the entity itself. More complex is adapting the radius to the context. The radius could be dependent on the current speed, larger and vaguer with higher speed, smaller and more precise with lower speed. Also an adaptation to the environment is possible, smaller radii in town bigger radii in rural regions.

5.4 Location-based services

This book has as its main focus the limited user interface availabilities of mobile phones. But with Flusi's architecture and the described techniques a very broad range of applications is imaginable. The ascertainable position information is a very valuable user information. Users can leave public information at their current position, readable and usable for every other user. The information can go from text to pictures, sound and video. Maps can be integrated and the application is augmented by a navigation solution. Also real satellite pictures, like those from Google Earth [30] are integrable over internet. Closed user groups can use buddy lists with information about the other members. So going out or meeting is possible without earlier arrangements. It's also just a small step from the Flusi architecture to so called "near by"-services and "yellow maps".

5.5 Payment

One possible categorization of Location-based services is into profit based and non-profit services. Profit-based services require secure payment and authentification and widely accepted and simple to use access mechanisms [31]. They are important drivers for the propagation of location-based services because they are powered by economic interests.

5.6 Protection of minors

Another aspect in public databases is the protection of minors. It has to be sure that in a public accessible database, juvenile endangering content is protected. A G-rated version could filter by a bad word-list X-rated content and offer only harmless topics.

5.7 Positioning

Through techniques like differential GPS a positioning augmentation is possible. Reference stations could broadcast correction values over the internet to augment the GPS position information of the mobile phone.

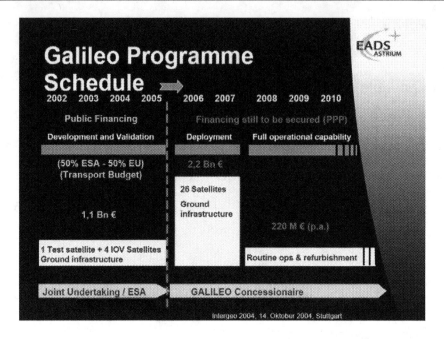

Figure 27: Galileo Programme Schedule [32]

Within the next years, the Navstar-GPS System will be expanded by the GALILEO programme (Figure 27). The currently 27 GPS satellites will be expanded by 30 GALILEO satellites. This entails a higher precision and a better availability. Because of the redundancy, positioning stays possible even if one system fails.

5.8 The future of user equipment

In future, the processing power, memory and connectivity of the used devices will increase significantly. Also the pervasiveness will increase. The user devices will be more multimodal. They will understand natural speak and gesture. Also the feedback will possibly be presented in more ways than graphical. So, devices have more possibilities to be "calm" and augmenting. With Wearable computers and the further development of smartphones, the dream of ubiquitous computers with location-based services will come true, with all the benefits and risks of these new techniques.

Appendix A

Overview about current Symbian cell-phones

Manufacturer	Available Model
Arima	U300
BenQ	BenQ P30
Foma	D901i
	F2051
	F2102V
	F700i
	F900i
Lenovo	P930
Motorola	Motorola A925
	Motorola A920
	Motorola A1000
	Motorola A1010
Nokia	Nokia 3230
	Nokia 3600
	Nokia 3650
	Nokia 3660
	Nokia 6600
	Nokia 6260
	Nokia 6620
	Nokia 6630
	Nokia 6680
	Nokia 6681
	Nokia 6682
	Nokia N-Gage
	Nokia N-Gage QD
	Nokia 3660/3620
	Nokia 7610
	Nokia 7650
	Nokia 7700
	Nokia 3650/3600
	Nokia 9210 Communicators
	Nokia 9290 Communicators
	Nokia 9300 Communicators
	Nokia 9500 Communicators

	Nokia N70
	Nokia N90
	Nokia N91
Panasonic	Panasonic X700
	Panasonic V800
Samsung	Samsung SGH-D710
Siemens	Siemens SX1
Sony Ericsson	Sony Ericsson P800
	Sony Ericsson P900
	Sony Ericsson P910i

Appendix B

Current location sensing technologies [20]

Technology	Technique	Physical	Symbolic	Absolute	Relative	LLC	Recognition	Accuracy and precision if available	Scale	Cost	Limitations
GPS	Radio time-of-flight lateration	•		•		✓		1-5 meters (95-99 percent)	24 satellites worldwide	Expensive infrastructure $100 receivers	Not indoors
Active Badges	Diffuse infrared cellular proximity		•	•			✓	Room size	1 base per room, badge per base per 10 sec	Administration costs, cheap tags and bases	Sunlight and fluorescent light interfere with infrared
Active Bats	Ultrasound time-of-flight lateration	•		•			✓	9 cm (95 percent)	1 base per 10 square meters, 25 computations per room per sec	Administration costs, cheap tags and sensors	Required ceiling sensor grid
MotionStar	Scene analysis, lateration	•		•			✓	1 mm, 1 ms, 0.1° (nearly 100 percent)	Controller per scene, 108 sensors per scene	Controlled scenes, expensive hardware	Control unit tether, precise installation
VHF Omini-directional Ranging	Angulation	•		•		✓		1° radial (≈100 percent)	Several transmitters per metropolitan area	Expensive infrastructure, inexpensive aircraft receivers	30-140 nautical miles, line of sight
Cricket	Proximity, lateration		•	○	○		✓	4 × 4 ft. regions (≈100 percent)	≈1 beacon per 16 square ft.	$10 beacons and receivers	No central management receiver computation
MSR RADAR	802.11 RF scene analysis and triangulation	•		•			✓	3-4.3 m (50 percent)	3 bases per floor	802.11 network installation, ≈$100 wireless NICs	Wireless NICs required
PinPoint 3D-iD	RF lateration	•		•			✓	1-3 m	Several bases per building	Infrastructure installation, expensive hardware	Proprietary, 802.11 interference
Avalanche Transceivers	Radio signal strength proximity	•			•			Variable, 60-80 meter range	1 transceiver per person	≈$200 per transceiver	Short radio range, unwanted signal attenuation
Easy Living	Vision, triangulation		•	•			✓	Variable	3 cameras per small room	Processing power, installation cameras	Ubiquitous public cameras
Smart Floor	Physical contact proximity	•		•			✓	Spacing of pressure sensors (100 percent)	Complete sensor grid per floor	Installation of sensor grid, creation of footfall training dataset	Recognition may not scale to large populations
Automatic ID systems	Proximity		•	○	○		✓	Range of sensing phenomenon (RFID typically <1m)	Sensor per location	Installation, variable hardware costs	Must know sensor locations
Wireless Andrew	802.11 proximity		•	•			✓	802.11 cell size, (≈ approx. 100 m indoor, 1 km free space)	Many bases per campus	802.11 deployment, ≈$100 wireless NICs	Wireless NICs required, RF cell geometries
E911	Triangulation	•		•			✓	150-300 m (95 percent)	Density of cellular infrastructure	Upgrading phone hardware or cell infrastructure	Only where cell coverage exists

Appendix C

DTD Universe

```
<?xml version="1.0" encoding="UTF-8"?>
<!ELEMENT Universe (Entity+)>
<!ELEMENT  Entity (URI, ELabell)>
     <!ATTLIST Entity Name CDATA #REQUIRED>
<!ELEMENT URI (#PCDATA)>
<!ELEMENT ELabel (#PCDATA)>
```

Appendix D

DTD Location

```
<?xml version="1.0" encoding="UTF-8"?>
<!ELEMENT Locations (Location+)>
     <!ATTLIST Location Determination (None | GPS | Time | CellID)  "None">
<!ELEMENT  Entity (URI, ELabell)>
<!ELEMENT Locname (#PCDATA)>
<!ELEMENT Locdata (#PCDATA)>
<!ELEMENT Locconnection (Entity?)>
<!ELEMENT Entity (#PCDATA)>
<!ELEMENT Loc (#PCDATA)>
     <!ATTLIST Loc Entities CDATA #REQUIRED>
```

Abbreviations

3G	Third Generation
API	Application Programming Interface
CPU	Central Processing Unit
DLL	Dynamic Link Library
DOM	Document Object Model
DOP	Dilution Of Precision
DTD	Document Type Definition
EDGE	Enhanced Data for GSM Evolution
FLUSI	Flexible User Interface
GNSS	Global Navigation Satellite System
GPRS	General Packet Radio Service
GPS	Global Positioning System
GSM	Global System Mobile
GUI	Graphical User Interface
HSCSD	High Speed Circuit Switched Data
HTML	HyperText Markup Language
HTTP	Hypertext Transfer Protocol
I/O Devices	Input / Output Devices
IT	Information Technologies
Navstar	Navigation System with Timing And Ranging
NMEA	National Marine Electronics Association
OO	Object Oriented
OS	Operating System
PC	Personal Computer
POI	Points Of Interest
RAM	Random Accessed Memory
ROM	Read Only Memory
SAX	Simple API for XML
U.S.	United States of America
UI	User Interface
UIQ	User Interface Quartz
UMTS	Universal Mobile Communication System
URI	Uniform Resource Locator
WGS 84	World Geodetic System of 1984
WLAN	Wireless Local Area Network
XML	eXtented Markup Language

Reference list

[1] Wikipedia. URL: http://www.wikipedia.org

[2] François Bergeron, Lin Gingras, Pierre Hadaya, Claude Caron. *A Framework for Evaluating StrategicLocation-based Applications in Businesses.* In Proceedings of the 38th Hawaii International Conference on System Sciences, Hawaii 2005.

[3] Symbian OS – the mobile operating system. URL: http://symbian.com

[4] Michael LYU Rung Tsong, Mok Ming Fai, Lee Kwok Chau. *Location-based services using GSM Cell Information over Symbian OS.* Final Year Project, Chinese University of Hong Kong, Hong Kong 2003 – 2004.

[5] Rosell, Peter; Kero, Roland. *Mobile Position Related Services with Personalization.* Master's Thesis, Lulea Tekniska Universitet, Lulea 1999.

[6] Ansgar Gerlicher, Stephan Rupp. *Symbian OS – Eine Einführung in die Anwedungsentwicklung.* Dpunkt Verlag GmbH; 2004.

[7] Richard Harrison. *Symbian OS C++ for Mobile Phones.* John Wiley & Sons Ltd; 2004.

[8] Jesper Andersson. *A Deployment System for Pervasive Computing.* Department of Computer and Information Science, Linköpings universitet, Sweden 2000.

[9] Series 60 Platform. URL:
http://www.series60.com/technicalinfo

[10] ARM. URL:http://arm.com

[11] Christian Lindholm, Turkka Keinonen, Harri Kilijander. *Mobile Usability.* McGraw-Hill Companies; 2003.

[12] Ubiquitous Computing. URL:
http://www.ubiq.com/hypertext/weiser/UbiHome.html

[13] Mark Weiser. *Ubiquitous Computing.* IEEE Computer "Hot Topics"; October 1993.

[14] Nishkam Ravi, Peter Stern, Niket Desai, and Liviu Iftode. *Accessing Ubiquitous Services Using Smart Phones.* Department of Computer Science, Rutgers University; USA 2004.

[15] William G. Griswold, Robert Boyer, Steven W. Brown, Tan
     Minh Truong. *A Component Architecture for an Extensible,*
     Highly Integrated Context-Aware Computing Infrastructure.
     Department of Computer Science and Engineering, University
     of California, San Diego 2003

[16] TA pervasive-computing. URL: http://www.ta-swiss.ch/www-
     remain/projects_archive/information_society/pervasive_d.htm

[17] Jochen Schiller. *Location based services.* Morgan Kaufmann
     Publishers, 2004.

[18] Leo Sauermann. *The Gnowsis.* Diplomarbeit, Vienna University
     of Technology, Vienna 2003.

[19] Matti Metsälä. *Positioning Framework for Symbian O.S.* Mas-
     ter's Thesis, Departement of Information Technology, Univer-
     sity of Turku, Finland 2004.

[20] Jeffrey Hightower, Gaetano Borriello. *Location Systems for*
     Ubiquitous Computing. IEEE Location Aware Computing, Au-
     gust 2001.

[21] Christian Becker, Frank Dürr. *On location models for ubiqui-*
     tous computing. Springer, Personal Ubiquitous Computing,
     September 2005.

[22] Tapio Pitkäranta. *Automatic and Semi-Automatic Context In-formation Extraction on Mobile Devices.* VTT Information Technology Finnland 2004.

[23] Jeffrey Hightower, Gaetano Borriello. *Location sensing tech-niques.* Computer Science and Engineering, University of Washington, USA 2001.

[24] Jeffrey Hightower, Gaetano Borriello, Lin Liao, Dirk Schulz. *Bayesian Filtering for location estimation.* IEEE Pervasive Computing, 2003.

[25] Bellocci V., Genovese, S., Inuggiato, D., Tucci,M. *Mobile Loca-tion-Aware Services: 2002 Market Perspective.* Ericsson, Divi-sion Service Architecture and Interactive Solutions, July 2002.

[26] Daniel Fallman. *From Freedom to involvement.* In Proceedings of the 38th Hawaii International Conference on System Sci-ences, Hawaii 2005.

[27] Gregory T. French. *Understanding the GPS.* GeoSearch Inc. 1996.

[28] Umts-forum.com. URL: http://www.umts-forum.com

[29] Hiroyuki Tarumi, Seiko Tokuda, Tomohiro Yasui, Kazuya Ma-
     tsubara, Fusako Kusunoki. *Design and Evaluation of a Loca-
     tion-Based Virtual City System for Mobile Phones.* In the 2005
     Symposium on Applications and the Internet 2005.

[30] Google Earth – Home. URL: http://earth.google.com

[31] Nishkam Ravi, Peter Stern, Niket Desai, Liviu Iftode. *Accessing
     Ubiquitous Services Using Smart Phones.* In Proceedings of the
     3rd IEEE Int'l Conference on Pervasive Computing and Commu-
     nications, 2005.

[32] GALILEO Stand des Programms und der Systemdefinition. URL:
     http://www.intergeo2004.de/rubrik/download/Schaefer.pdf